Illustrations by Mahzar Iqbal.

About the author.

The author is a teacher, writer and independent researcher.

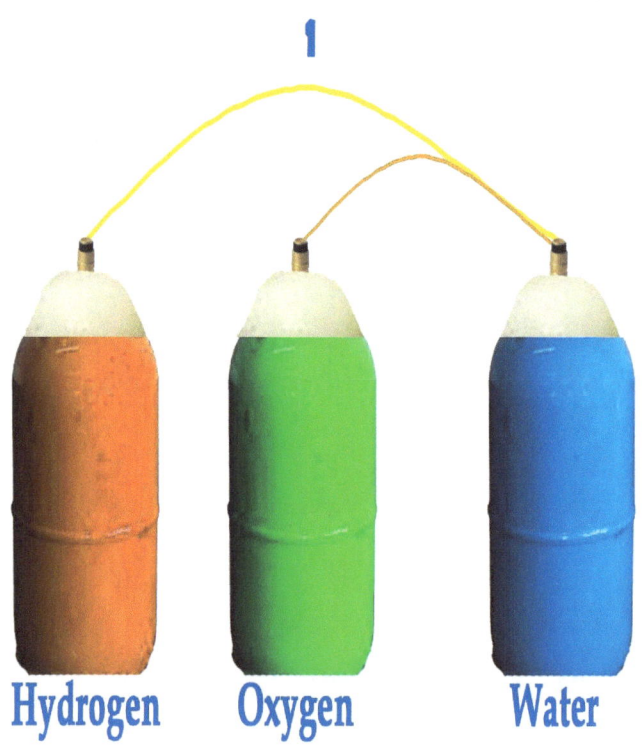

Hydrogen   **Oxygen**   **Water**

We are going to make water

2

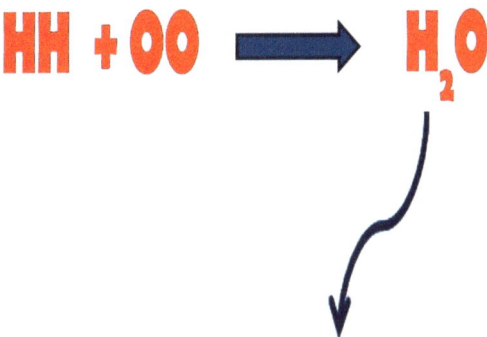

$HH + OO \longrightarrow H_2O$

Here is some problem!
Why it is not like this
H_2O_2

Fact: H_2O is Water
H_2O_2 is poison

3

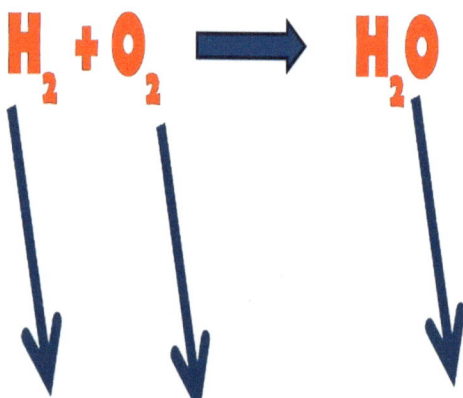

$$H_2 + O_2 \longrightarrow H_2O$$

Two Atoms of Hydrogen

Two Atoms of Oxygen

Water molecule after the arrow, the atoms are not same as before the arrow

4

You have to follow a Law

Law???   (I know, you are not going to be
                  a lawyer).
Ok, What is that law?

Law of Conversation of Mass.
    "Mass can neither be created nor be destroyed"

5

Let's apply this law with our problem

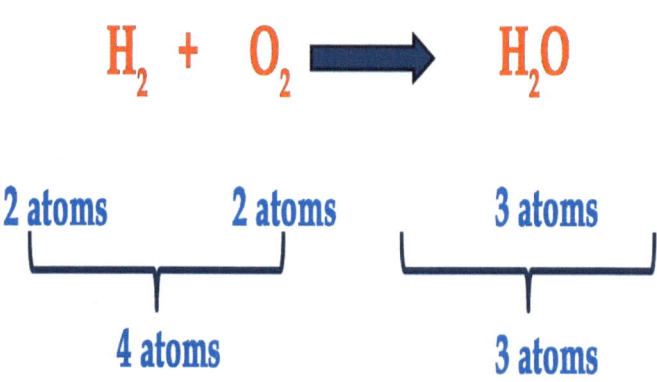

$$H_2 + O_2 \longrightarrow H_2O$$

2 atoms   2 atoms   3 atoms

4 atoms   3 atoms

* More atoms on the left hand side and
less atoms on the right hand side.

Rrecall the law, mass (atoms) can neither be created nor be destroyed.

Let's Look at our equation

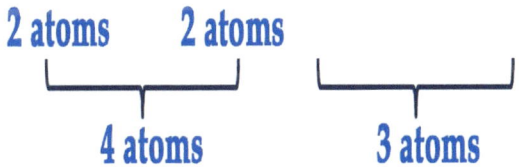

$$H_2 + O_2 \longrightarrow H_2O$$

2 atoms   2 atoms

4 atoms   3 atoms

Hmm!!!! we have to balance it.

Ok, think and start

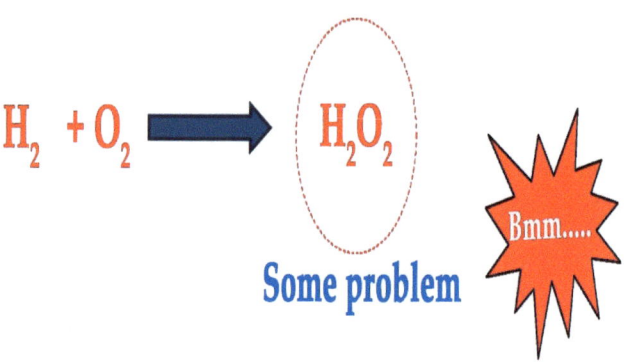

$$H_2 + O_2 \longrightarrow H_2O_2$$

Some problem

Bmm.....

H_2O_2 is hydrogen peroxide, a poison,
we can not change the formula of water molecule
It is always H_2O.

So, how to cater this problem
Let's use little bit of Maths

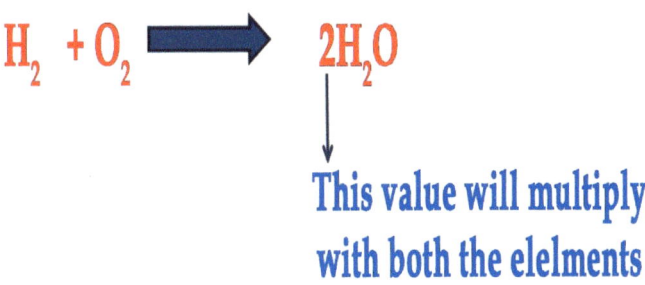

$$H_2 + O_2 \longrightarrow 2H_2O$$

This value will multiply
with both the elelments

$2 \times H_2 \quad O$

So we get 4 H atoms and 2 oxygen atoms.

Let us count the atoms again

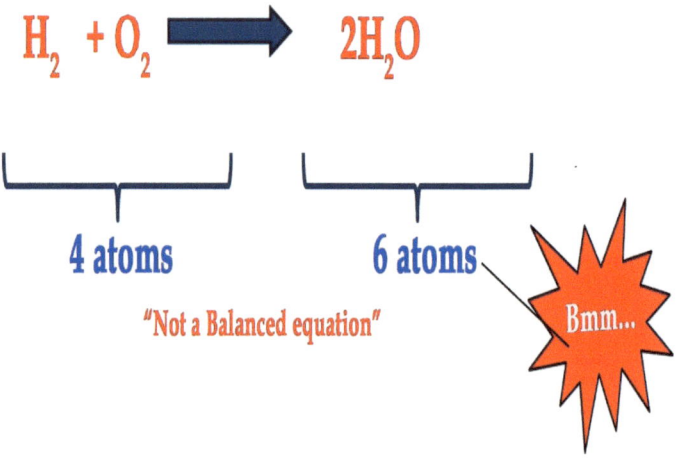

$$H_2 + O_2 \longrightarrow 2H_2O$$

4 atoms   6 atoms

"Not a Balanced equation"

Bmm...

Its really confusing !!
But I am not going to budge.

Doing a little maths on the left hand side

$$2H_2 + O_2 \longrightarrow 2H_2O$$

6 atoms   6 atoms

Wow!!!   Its Balanced

I think, I am next to Einstein

11

Easy Questions about balancing:

Q: What is the Formula of Water?

Q: What are reactants?

Q: What are Products?

Q:What is the law of conversation of mass?

Let's go for another example

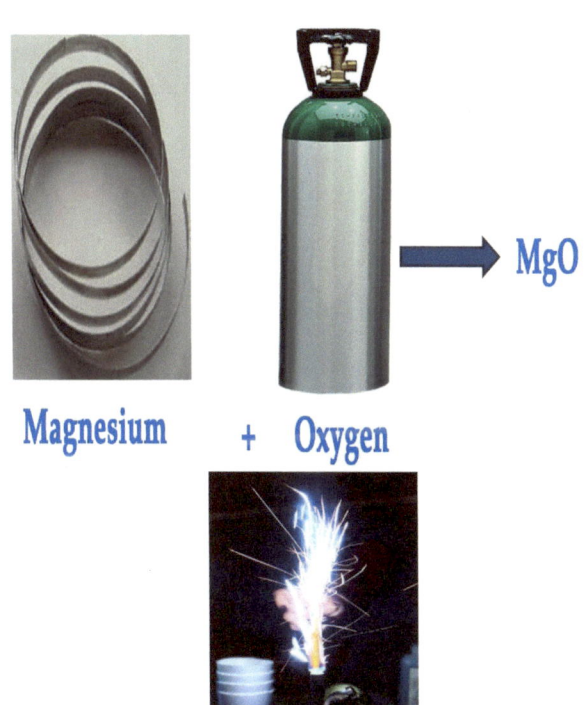

Magnesium   +   Oxygen   ➜   MgO

Let's Balance this Equation:

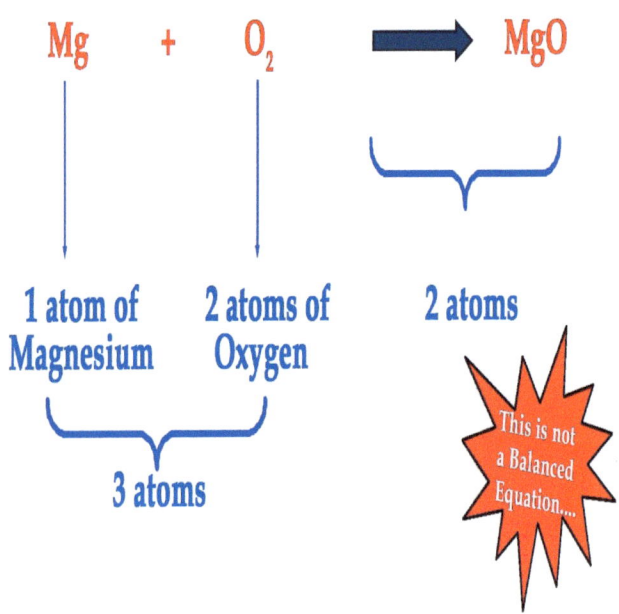

Mg + O_2 ⟶ MgO

1 atom of Magnesium

2 atoms of Oxygen

2 atoms

3 atoms

This is not a Balanced Equation....

14

Think, guess, Predict ! ! !

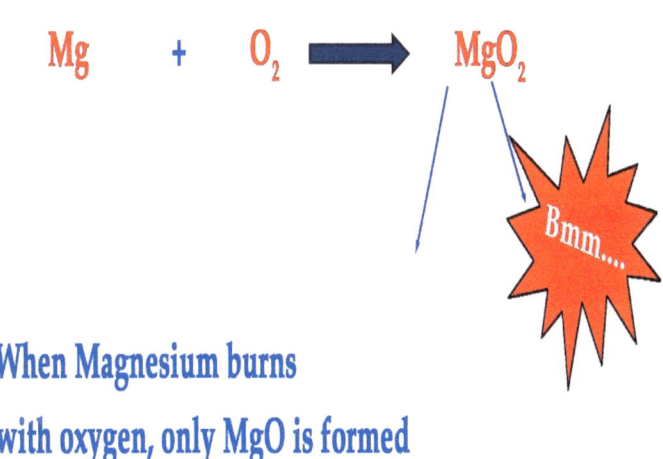

$$Mg \quad + \quad O_2 \quad \longrightarrow \quad MgO_2$$

Bmm....

When Magnesium burns
with oxygen, only MgO is formed
we can not change the formula...

Think Mathematically

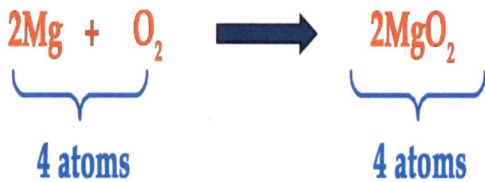

$$2Mg + O_2 \longrightarrow 2MgO_2$$

4 atoms   4 atoms

Here we go

That's great you are Davinci of this world ! !

16

Tips + Tricks for Balancing

* Practice simple equations

* Learn formula of common compounds

*Do more and more lab activity

Being good at symbols means you have more
More Know now of elements

Some Common Elements

Elements	Symbols
Hydrogen	H
Sodium	Na
Potassium	K
Magnesium	Mg

Elements	Symbol
Calcium	Ca
Zine	Zn
Copper	Cu
Lead	Pb
Sulphur	S
Carbon	C

19

Understanding the formula of some common compounds.

Hydrogen

H_2

20

At times, we have to follow rules rather than our liking

Hydrogen (H) exists in nature as a Diatomic Molecule (H) ----- (H)

So, by rules we have to write formula of hydrogen as H_2

Same in the case for oxygen (O_2)
Chlorine (Cl_2) and Nitrogen N_2

* Every one is aware of the formula of water H_2O

* A very important acid hydrochloric acid is
  mainly used in every school lab

It has a simple formula Hcl

* Suffocating gas carbon monomide, CO comes out during burning process.

* Co_2 Carbon dioxide is present in the air with a combination of carbon and two oxygen atoms.

* This form changes to Co_3 when it combines with calcium $CaCO_3$.

* $Caco_3$ calcium carbornate is marble used as flooring

* Carbondioxide is given during burning process

24

Other Book by the same Author

MAZHAR IQBAL AUTHOR

www.ingramcontent.com/pod-product-compliance
Lightning Source LLC
Chambersburg PA
CBHW040814200526
45159CB00022B/915

* 9 7 8 1 5 3 0 2 0 7 9 5 4 *